The Iron Bridge

Ironbridge, England

Iron Bridge, 2019,
Luftaufnahme von Südosten / aerial view from the south-east

Iron Bridge, 2019,
Nordost-Ansicht / view from the north-east

Iron Bridge, 2019,
Blick vom südlichen Ufer
auf den gleichnamigen
Ort mit der St.-Lukas-
Kirche von 1837 /
the town bearing its name
and St Luke's Church
built in 1837 viewed from
the south bank

Project data

Iron Bridge, 1777–1779
Constructed by: Thomas Farnolls Pritchard, John Wilkinson, Abraham Darby III
Total length: 60 metres (197 ft), span: 30.62 metres (100 ft 6 in), height of the arches: 12.80 metres (42 ft)
Amount of cast iron used: 385 tonnes

Restoration:
2017–2019

Owner:
Telford & Wrekin Council

Commissioned by:
English Heritage, London

Architects / structural engineers:
Ferguson Mann Architects, Bristol; The Morton Partnership, London

Contractor:
Taziker Industrial Ltd., Bolton

Costs:
4,200,000 euros

Funding provided by HERMANN REEMTSMA STIFTUNG:
1,000,000 euros

Address:
The Iron Bridge, Ironbridge, Telford, Shropshire, TF8 7JP, United Kingdom

Opening hours:
Open to the public all year round.

Special Features:
The world's oldest cast iron arch bridge.
First notable structure of the modern industrial age.
Planned and realised by a private business consortium.
UNESCO World Heritage Site.

Fakten zum Bauprojekt

Iron Bridge, 1777–1779
Erbauer: Thomas Farnolls Pritchard, John Wilkinson, Abraham Darby III.
Gesamtlänge: 60 Meter, Spannweite: 30,62 Meter, Bögenhöhe: 12,80 Meter
Verbautes Gusseisen: 385 Tonnen

Restaurierung:
2017–2019

Eigentümer:
Telford & Wrekin Council

Bauherr:
English Heritage, London

Architekten / Bauingenieure:
Ferguson Mann Architects, Bristol; The Morton Partnership, London

Ausführende Firma:
Taziker Industrial Ltd., Bolton

Kosten:
4.200.000 Euro

Förderung der HERMANN REEMTSMA STIFTUNG:
1.000.000 Euro

Adresse:
The Iron Bridge, Ironbridge, Telford, Shropshire, TF8 7JP, Vereinigtes Königreich

Öffnungszeiten:
Ganzjährig öffentlich zugänglich.

Besonderheit:
Älteste gusseiserne Bogenbrücke der Welt.
Erstes markantes Bauwerk der modernen Industriegeschichte.
Geplant und realisiert durch ein privates Unternehmerkonsortium.
UNESCO-Welterbe.

Blick auf die Gusseisenbrücke bei Coalbrookdale /
View of the cast iron bridge near Coalbrookdale,
Kunstdruck von / fine art print by William Ellis,
(nach einem Gemälde von / after a painting by Michael »Angelo« Rooker), London 1782

Entstehungsgeschichte, Standort und Bauaufgabe

Etwa 60 km nordwestlich von Birmingham, in der Grafschaft Shropshire, liegen im bewaldeten Tal des Flusses Severn die heute zur Stadt Telford gehörenden Ortschaften Ironbridge und Coalbrookdale. Die Gegend wird als »birthplace of the Industrial Revolution« bezeichnet. Die Ortsnamen weisen auf die zwei wesentlichen Faktoren dieser Wiege der industriellen Revolution hin: Eisen und Kohle. Die geologische und topografische Situation war günstig. Nach Ende der letzten Eiszeit entstand vor etwa 15.000 Jahren die Schlucht (gorge) des Severn und brachte große Steinkohle-, Eisenerz-, Ton- und Kalkvorkommen hervor. Seit der Zeit von Heinrich VIII. wurde dort Eisenerz gefördert und mit Holzkohle zu Schmiedeeisen verarbeitet. Im 17. Jahrhundert verwandelte sich die romantische Hügellandschaft zunehmend in eine frühindustrielle Fertigungsstätte. Dem Eisenhüttenunternehmer Abraham Darby (1676–1717) gelang es 1709 erstmals, brauchbares Roheisen mit Koks zu gewinnen. Koks ist kohlenstoffreich, beim Verbrennen fällt im Vergleich zu Kohle weniger Rauch, Ruß und Schwefel an. Darby erzeugte das Koks aus schwefelarmer Kohle der naheliegenden Flöze bei Wärmezufuhr und unter Sauerstoffabschluss. Das Gusseisen entstand beim Schmelzen des aufbereiteten Eisenerzes in Hochöfen bei etwa 1.150 Grad Celsius. Dieses Roheisen konnte nicht geschmiedet werden, da die Schwefelanteile noch höher sind als beim Schmiedeeisen. Es war hart, spröde, druckstabil und korrosionsbeständig. Es konnte verhältnismäßig einfach in großen Mengen hergestellt werden und ließ sich – da die Schmelze dünnflüssig ist – in Sandformen gießen. Auf diese Weise konnten alle erdenklichen Eisenteile hergestellt werden. Mit Darbys neuer Methode war die Produktion von Gusseisen unabhängig von Holz möglich, das in England im 18. Jahrhundert knapp wurde. Darby baute neue und

größere Öfen – sein Betrieb, die Coalbrookdale Company, wuchs.
Die neue Technologie war sowohl ein wesentlicher Ausgangspunkt für Englands Führungsrolle als Industrienation, als auch Grundlage für weitere Erfindungen, wie das Puddelverfahren (Umwandlung von Roheisen in Schmiedeeisen) und schließlich die Gewinnung von Stahl, die bis heute gängigste Eisen-Kohlenstoff-Legierung.
Abraham Darby entstammte einer Quäkerfamilie. Diese religiöse Gemeinschaft hatte sich Mitte des 17. Jahrhunderts in Nordwestengland gebildet, sie stand für tugendhaftes Leben und die Entfaltung individueller Fähigkeiten. Sein Sohn Abraham Darby II. (1711–1763) führte das Unternehmen weiter und tat sich ebenfalls als Erfinder von Verfahren zur Verbesserung der Eisengewinnung und -verarbeitung hervor. Abraham Darby III. (1750–1791) übernahm mit 18 Jahren die Geschäftsleitung der Coalbrookdale Company, die sowohl Räder und Kessel für große Dampfmaschinen, als auch Töpfe für den häuslichen Bedarf herstellte.
Der Abtransport der Eisenprodukte verlief über den Severn. An beiden Seiten des Flusses hatten sich Förder- und Produktionsstätten angesiedelt, es gab Fährbetriebe und Holzbrücken. Doch jahreszeitenbedingtes Hoch- und Niedrigwasser schränkte den zuverlässigen Personen- und Warenverkehr ein.
Thomas Farnolls Pritchard (1723–1777), ein erfolgreicher Architekt der West Midlands, beschäftigte sich schon länger mit Brückenkonstruktionen, als er 1773 seinem Freund und »Ironmaster« John Wilkinson (1728–1808) Pläne für eine Brücke über den Severn zwischen Coalbrookdale und Broseley vorstellte. Wilkinson war zum Ende des 18. Jahrhunderts der angesehenste Fachmann auf dem Gebiet des Eisenhüttenwesens in England, sein Spitzname lautete »Iron-Mad Wilkinson«.

Er war Erfinder, Unternehmer und überzeugter Eisenfabrikant, der mit spektakulären Konstruktionen und innovativen Projekten die neuen Möglichkeiten des Materials demonstrierte – 1787 ließ er das erste Boot aus Eisen zu Wasser.

Thomas Farnolls Pritchard,
unbekannter Künstler / unknown artist,
vor / before 1777

History, location and purpose

Around sixty kilometres to the north of Birmingham, the towns of Ironbridge and Coalbrookdale nestle in a wooded gorge cut through by the River Severn. Situated on the outskirts of Telford in the county of Shropshire, they are known as the 'birthplace of the Industrial Revolution'. Their names point to two essential ingredients that allowed local industrial production to flourish: iron and coal. The area's favourable geological and topographical conditions were created around 15,000 years ago towards the end of the last ice age when the Severn Gorge was formed, exposing vast layers of hard coal, iron ore, clay and limestone. Iron ore had been mined in the region as far back as the times of Henry VIII and was smelted using charcoal. During the seventeenth century, the picturesque hills and dales were gradually transformed into an early centre of industrial activity. In 1709, the ironmaster Abraham Darby (1676–1717) succeeded in producing marketable pig iron by using coke as a fuel. Coke is rich in carbon and emits less smoke, soot and sulphur during smelting. Darby used low-sulphur coal extracted from nearby deposits which he heated out of contact with air, producing cast iron by smelting iron ore in a blast furnace at a temperature of roughly 1,150 degrees centigrade. The resulting pig iron was not malleable as it still had a higher sulphur content than wrought iron. It was hard, brittle and resistant to pressure and corrosion. It was relatively easy to produce in large quantities and could be poured into moulds formed in the sand owing to the low viscosity of the molten metal. With this method, many different iron products could be manufactured. Darby's new technique allowed cast iron to be made without using wood, which was becoming a scarce resource in eighteenth-century England. His Coalbrookdale Company grew as he kept building better and larger furnaces.

Not only was this technological innovation an essential milestone in England's rise as a leading industrial nation, but it also paved the way for further advances in metallurgy such as puddling (converting pig iron into wrought iron) and ultimately steel production, which remains the most widely used alloy of iron and carbon to this day.

Abraham Darby came from a Quaker family, a religious community that formed in the north-west of England around the middle of the seventeenth century and stood for a virtuous life and the cultivation of individual abilities. His son Abraham Darby II (1711–1763) took over his father's business and similarly continued to refine the methods of iron production and processing. At the age of eighteen, Abraham Darby III (1750–1791) became the director of the Coalbrookdale Company, which manufactured wheels and boilers for large steam engines as well as pots for domestic use.

The iron products were transported on the River Severn. Mines and factories had been established along both sides of the river, which were linked by ferries and wooden bridges. However, seasonal floods and periods of low water regularly caused disruptions to the transport of people and goods.

Thomas Farnolls Pritchard (1723–1777), an acclaimed West Midlands architect who had some experience in bridge construction, drafted plans for a bridge across the Severn between Coalbrookdale and Broseley. In 1773, he showed the plans to his friend, the ironmaster John Wilkinson (1728–1808). Wilkinson, or 'Iron-Mad Wilkinson' as he was known, was the most renowned expert in iron manufacturing in late eighteenth-century England. An inventor, industrialist and dedicated ironmaster, he devised spectacular constructions and innovative projects to demonstrate new applications for the material, for instance when he launched the first iron barge in 1787.

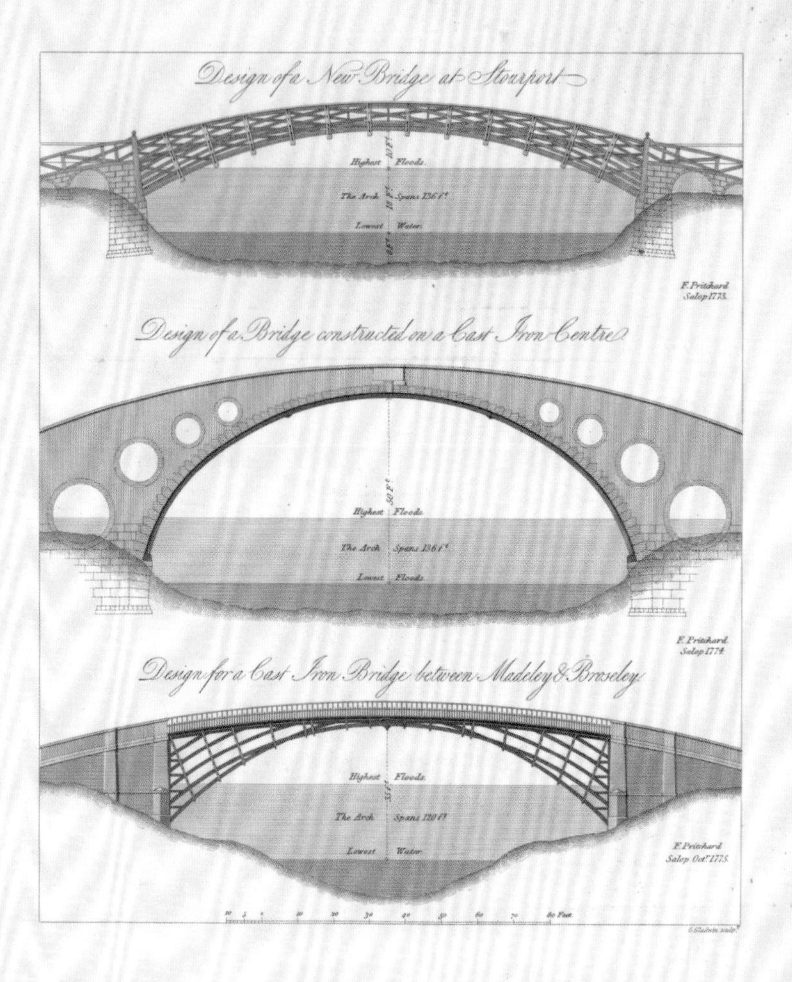

Design of a New Bridge at Stourport
Highest Floods.
The Arch Spans 136 F.t
Lowest Water.
F. Pritchard
Salop 1773.
Design of a Bridge constructed on a Cast Iron Centre
50 F.t
Highest Floods.
The Arch Spans 136 F.t
Lowest Floods.
F. Pritchard
Salop 1774.
Design for a Cast Iron Bridge between Madeley & Broseley
Highest Floods.
35 F.t
The Arch Spans 120 F.t
Lowest Water.
F. Pritchard
Salop Oct.r 1775.
10 5 0 10 20 30 40 50 60 70 80 Feet

←
Die drei Entwurfsstadien der Iron Bridge /
The three design stages of the Iron Bridge
Auszug aus / Excerpt from John White,
On Cementitious Architecture as applicable to the Construction of Bridges, London 1832

Der Bau der Iron Bridge / The Iron Bridge under construction,
Aquarell von / watercolour sketch by Elias Martin, Juli / July 1779,
einzig bekannte Darstellung der Iron Bridge während des Baus /
only known painting showing the Iron Bridge during construction

William Williams, *The Iron Bridge*,
Gemälde (Ausschnitt) / painting (detail), 1780

Details der Iron Bridge in Coalbrookdale / Details of the Iron Bridge in Coalbrookdale,
Kupferstich von / copperplate engraving by John Record,
herausgegeben von / edited by James Phillips, 1782

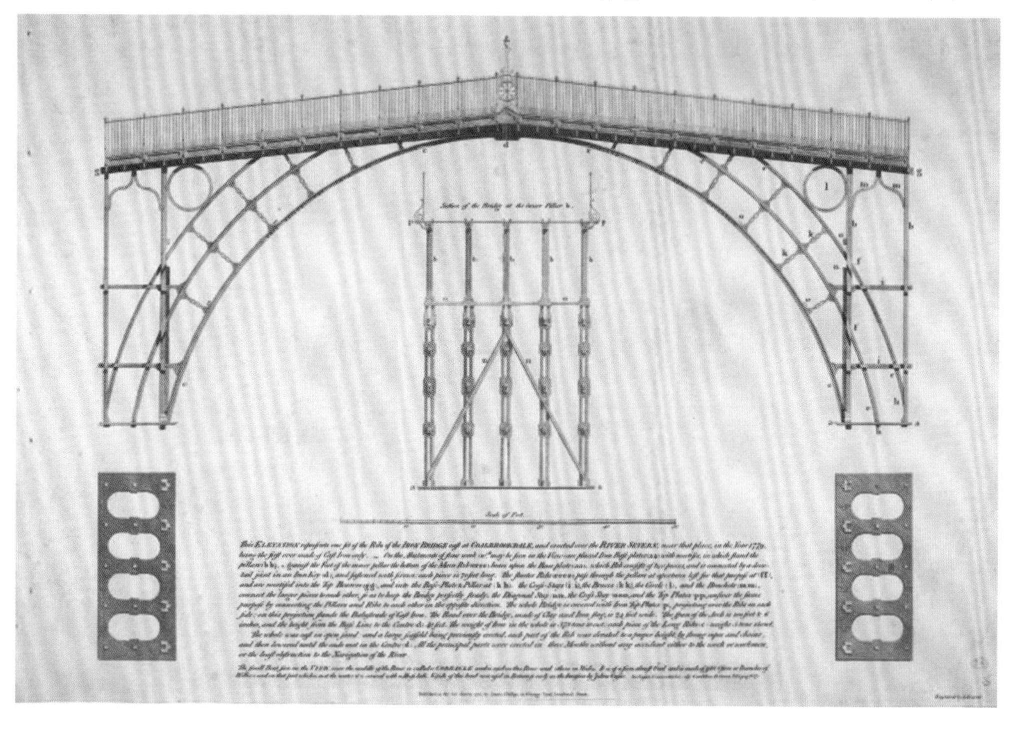

Entwurf, Konstruktion und Ausführung

John Wilkinson unterstützte das Iron-Bridge-Projekt nachdrücklich und beriet Architekt Pritchard in technischen Fragen. Der entwarf die Brücke in drei Entwicklungsschritten, wie eine Publikation von John White zeigt: zunächst eine Variante aus Holz, eine gemauerte auf einem Eisenkern, schließlich die erste Version einer Eisenbogenbrücke mit zwei Auflagern aus Stein. Mit diesem Plan wurden Unterstützer und Unternehmer im Severntal angeworben, die man vor allem um Abraham Darby III. fand. Er bildete federführend ein Konsortium zur Finanzierung des Projekts. Im Oktober 1775 wurde beim Parlament in London ein Budgetplan zur Genehmigung eingereicht, die Kalkulation ging von Gesamtkosten in Höhe von 3.200 Pfund aus. Als Platz für die Brücke wurde eine Fährüberquerung mit Straßenanschluss am Fuße des Ortes Broseley ausgewählt. Im November 1777 begannen die Fundamentarbeiten für die Auflager. Abraham Darby III. hatte inzwischen die Leitung des Vorhabens übernommen, Thomas Farnolls Pritchard war gestorben, John Wilkinson hatte seine Anteile verkauft.

Für die Konstruktion der Iron Bridge gab es keine Vorbilder, man konnte auf keine Berechnungen oder Erfahrungswerte zur Belastbarkeit großer Eisengussteile zurückgreifen. So folgt die Iron Bridge dem Prinzip hölzerner Binderkonstruktionen. Die schließlich ausgeführte Version sah fünf Trägerrippen vor, die im Abstand von etwa 1,50 Meter nebeneinandergesetzt wurden. Sie bestehen jeweils aus drei konzentrischen Bögen, die mit radialen Sprossen verbunden sind. Die Bögen sind wiederum mit seitlichen Ständern verstrebt, die eine Verbindung zu den Auflagern schaffen. Große Eisenringe und Kielbögen in den Zwickeln deuten auf Pritchard als Urheber hin, der solche Ornamente auch in anderen Bauwerken einsetzte. Der Ring wurde in späteren Brücken ein oft aufgegriffenes Motiv. Der größte Bogen hat eine Spannweite von 30,62 Metern. Er ist aus zwei je 21,60 Meter langen Halbbögen zusammengesetzt, die jeweils über 5 Tonnen wiegen. Aufgestellt ergeben sie eine Bogenhöhe von etwa 13 Metern. Insgesamt sollen 385 Tonnen Gusseisen und über 1.700 Einzelteile verbaut worden sein. Es kamen keine Bolzen oder Nieten, sondern ausschließlich zimmermannsmäßige Zapf- und Steckverbindungen zum Einsatz.

Die ganze Konstruktion stand auf dicken Eisenplatten, die mit dem Fundament aus Stein unterfangen waren. Das Pfeilermauerwerk aus regionalem Sandstein links und rechts am Ufer ist wahrscheinlich erst nach dem Errichten der Eisenkonstruktion angebaut worden, mit kleinen Torbögen zur Durchfahrt längs des Flusses. Die massiven Auflager geben gestalterisch den Widerpart zur schlanken Eisenkonstruktion. Die etwa 7,30 Meter breite Fahrbahn, die von einem formschönen hohen Geländer aus Gusseisen eingefasst ist, führt leicht ansteigend zum Scheitelpunkt des Brückenbogens.

Sie bestand ursprünglich aus gusseisernen Platten, die auf den Trägern ruhten, und einer Decklage aus Ton und Kohlenschlacke.

Der genaue Produktionsprozess und -ort der Eisenteile, der Transportweg zur Baustelle und die Art der Montage sind nicht dokumentiert. Man geht aber davon aus, dass das Eisen in Darbys Hochöfen in Coalbrookdale produziert wurde. Die Gussformen entstanden, indem man Modelle aus Holz in Sand drückte. Eine historische Zeichnung von der Baustelle zeigt, dass die Bögen mithilfe eines besonderen Flaschenzuges auf einem Boot herbeigeschifft wurden. Die Aufstellung der Brücke erfolgte im Sommer 1779.

In die beiden Abschlussbögen ist die deutlich sichtbare Inschrift eingegossen worden: »This bridge was cast at Coalbrook-Dale and erected in the year MDCCLXXIX«.

Planning, design and construction

John Wilkinson strongly supported the Iron Bridge project and assisted Pritchard with technical advice. As a book published by John White testifies, the achitect's draft proposed three developmental stages: first a wooden version, then a brick version erected on an iron core and finally a prototype of an iron arch bridge resting on two stone abutments. The design was used to enlist the support of sponsors and entrepreneurs from the Severn Gorge, who were mostly associated with Abraham Darby III. A consortium was put together with him at the helm to finance the project. In October 1775, they petitioned Parliament for approval of their budget plan, which put the total cost at £ 3,200. The site chosen for the bridge was a ferry crossing with an access road at the foot of the town of Broseley. The foundations of the abutments were laid in November 1777. By then, Abraham Darby III had been commissioned to carry out the project, Thomas Pritchard had died and John Wilkinson had sold his shares.

The construction of the Iron Bridge was without precedent; nobody had any experience with, or had ever calculated, the load-bearing capacity of large iron castings. As a result, the bridge was assembled using traditional timber construction techniques. The final structure is supported by five parallel ribs spaced around 1.5 metres (5 ft) apart. Each rib is composed of three concentric arches braced by radial struts. The arches are in turn attached to a vertical frame on either side of the river that connects them to the abutments. The large iron circles and ogees in the spandrels suggest that the design was Thomas Pritchard's, who also used these decorative elements in other structures. The circle became a characteristic design feature of many subsequent bridges. The largest rib has a span of 30.62 metres (100 ft 6 in) and is composed of two half-arches, each 21.6 metres (71 ft) long and weighing more than 5 tonnes. Standing upright, they have a height of around 13 metres (42 ft). In total, more than 1,700 individual pieces and 385 tonnes of cast iron are thought to have been used. Instead of bolts and rivets, the bridge is held together only by woodworking joints such as mortises and tenons, dovetails and wedges.

The structure was erected on heavy iron plates underpinned by a stone foundation. Local sandstone was used for the masonry of the piers on either side of the river. Forming an archway to accommodate a towpath along the river, the piers were probably added after the iron structure had been put in place. The massive abutments contrast markedly with the slender ironwork. The roadway, which has a width of around 7.3 metres (24 ft) and is bordered by tall, elegant cast iron railings, rises at a slight gradient on both sides towards the centre of the bridge. Originally, the deck consisted of cast iron plates resting on supports and a surface layer made of clay and slag.

Nothing is known about exactly where and how the iron components were produced, how they were transported to the construction site and how they were assembled, but it is generally thought that the ironwork was cast in Abraham Darby's furnaces in Coalbrookdale. The moulds were made by pressing wooden forms into the sand.

A historical drawing of the construction site shows that the arches were transported by boat using a special type of pulley. The structure was assembled in the summer of 1779. The two outer ribs are marked with the clearly visible inscription 'This Bridge was cast at Coalbrook-Dale and erected in the year MDCCLXXIX'.

→

Iron Bridge, 2019,
großer Gusseisenbogen, Radialsprossen,
eiserne Fahrbahnträger, Brüstung und Geländer /
large iron arch, radials, iron bridge deck,
parapet and railings

Iron Bridge, 2019,
Nordwest-Ansicht, vorne der Auflagerbau aus Stein, auf der südlichen Seite hinten schließen sich zwei kleine gusseiserne Bögen an, die später hinzugefügt wurden /
view from the north-west with the stone abutment in the foreground and two small iron arches in the background which were added later on the south side

COALBROOK=DALE

Iron Bridge, 2019,
Spannbögen, die mit den Radialsprossen verzapft sind,
Eisenringe, die die Bögen mit den seitlichen Ständern verstreben /
ribs braced by radials using mortises and tenons, iron circles
connecting the ribs to the lateral frame

Iron Bridge, 2019, Scheitelpunkt des Brüstungsgeländers mit dem eingegossenen Errichtungsdatum 1779 / central section of the parapet railings inscribed with the year of construction 1779

Iron Bridge, 2019,
Blick von unten auf die fünf Spannbögen, Verstrebungen und Bodenplatten /
the five braced ribs and the bridge deck viewed from below

Nutzung, Rezeption und Restaurierung

Die Iron Bridge wurde am Neujahrstag 1781 für den Verkehr geöffnet, acht Jahre nach den ersten Plänen von Pritchard. Auf der Südseite wurden die Zufahrtsstraßen und ein Zollgebäude fertiggestellt. Ein Dekret des Parlaments hatte die Brücke als Privateigentum und die Erhebung von Zoll für jedermann – auch für das Militär und die Krone – durch die Treuhänder bestätigt. Zwar hatten sich die Baukosten, die Abraham Darby III. hauptsächlich trug, verdoppelt, aber langfristig stellte sich das Projekt als unternehmerischer Erfolg heraus. Die Zolleinnahmen sicherten den Anteilseignern eine jährliche Rendite von acht Prozent.

Rasch entwickelte sich die Brücke zur Hauptquerung des Severn. An der Nordseite entstand eine Siedlung, die später den Ortsnamen Ironbridge erhielt. Die Brücke wurde durch Berichte, Bilder und Drucke überregional bekannt und lockte viele auswärtige Besucher an. Ein verkleinerter Nachbau wurde bereits 1791 im Maßstab 1:4 als Fußgängerbrücke im Wörlitzer Park in Deutschland errichtet – die erste Gusseisenbrücke auf dem europäischen Festland. Weitere Brückenprojekte nach dem Vorbild der Iron Bridge entstanden über den Severn in Buildwas (1796) und Coalport (1818), in Wearmouth im Nordosten Englands (1796) und im schottischen Craigellachie (1814). Die Iron Bridge revolutionierte nicht nur den Brückenbau. Die nun herstellbaren großen Eisenteile ermöglichten große Spannweiten und die Skelettbauweise und damit Markthallen, Bahnhöfe, den berühmten Crystal Palace und Wolkenkratzer, die eine ganze Architekturepoche ausmachten.

Durch geologische Verschiebungen im steilen Severntal traten in den Folgejahren immer wieder Risse in den Widerlagerbauten auf, auch mussten die Hänge mit Stützmauern abgefangen werden. Neben der Begeisterung für das neue Bauwerk kam Kritik auf,

die erst 1795 verstummte, als die Iron Bridge als einzige Severn-Brücke ein zerstörerisches Hochwasser überstand, da das Wasser durch die offene Eisenkonstruktion fluten konnte. Nach erheblichen Schäden an den südlichen Steinbögen wurden sie durch Holzkonstruktionen und 1820 durch Bögen aus Gusseisen ersetzt.
1934 musste die Brücke wegen starker Belastung für den Autoverkehr gesperrt werden, in diesem Jahr erfolgte auch die Anerkennung als nationales Denkmal. Zoll für Fußgänger wurde nur noch bis 1950 erhoben, als die Brücke ins Eigentum des Shropshire County Council (heute Telford & Wrekin Council) übertragen wurde.
1973–1975 setzte man unter der Brücke einen umgekehrten Bogen aus Stahlbeton ins Flussbett, um den nach innen strebenden Kräften der beiden Auflager entgegenzuwirken, deren Fundamente durch Hochwasser immer wieder unterspült wurden. Zudem versteifte man die südlichen Steinpfeiler mit Beton und ersetzte den Brückenbelag durch einen leichteren Asphalt.
1986 wurde die Brücke mit dem gesamten Tal (»Ironbridge Gorge«) als UNESCO-Welterbe eingestuft.
Die originale Gusseisen-Konstruktion blieb über die Zeit weitgehend unberührt. Sie erhielt lediglich neue Anstriche sowie zusätzliche Metall-Klammern und Bolzen, um die Zapf- und Steckverbindungen zu stabilisieren, die bei auftretenden Zugkräften gerissen waren. Die Denkmalorganisation English Heritage, in deren Obhut die Iron Bridge kam, führte zwischen 1999 und 2000 eine gründliche Begutachtung durch. Weitere intensive Forschungen zu Materialbeschaffenheit, Oberfläche, Konstruktion, aber auch zur Brückengeschichte schlossen sich an, ein detailliertes Konservierungskonzept wurde in Zusammenarbeit mit der Denkmalbehörde Historic England erarbei-

tet. Es stellte sich heraus, dass die gesamte Eisenkonstruktion durch die geologischen Verschiebungen und ein Erdbeben gegen Ende des 19. Jahrhunderts mehr geschädigt war als angenommen.
Nach Plänen von Ferguson Mann Architects aus Bristol und den Fachingenieuren The Morton Partnership aus London wurde die auf Brückensanierungen spezialisierte Firma Taziker Industrial aus Bolton mit den Arbeiten beauftragt. English Heritage formulierte die Ziele der Maßnahme: Erhalt der originalen Bausubstanz, Verhinderung von weiteren Verlusten von historischem Material, möglichst minimale Eingriffe, pragmatisches Vorgehen und bewährte technische Lösungen bei der Behebung von Mängeln.
Das Konservierungsprojekt begleitete English Heritage mit einer Vermittlungs- und Beteiligungskampagne. Man setzte auf eine intensive Medieninformation, startete eine Crowdfunding-Aktion und ermöglichte besondere Einblicke (»Conservation in Action«). Dafür wurde ein Laufsteg in das Baugerüst integriert, von dem aus Besucher den Handwerkern und Restauratoren bei der Arbeit zuschauen konnten und Erläuterungen von fachkundigen Freiwilligen erhielten. Die HERMANN REEMTSMA STIFTUNG war Förderpartner und brachte mit einer erheblichen Zuwendung den Durchbruch bei der Finanzierung und damit das Projekt in Gang.

Use, public response and restoration

The Iron Bridge opened to traffic on New Year's Day 1781, eight years after Pritchard had proposed his initial designs. Access roads and a toll house were built on the south side. A decree issued by Parliament had confirmed that the bridge was private property and that the trustees were permitted to collect tolls from the general public and even from soldiers and the Royal Family. Although the construction costs – most of which were paid for by Abraham Darby III – had doubled, the project ultimately proved a commercial success. Shareholders received an annual dividend of eight per cent from the tolls.

The bridge quickly became the most important Severn crossing. A settlement developed on the north side, which was later baptised Ironbridge. Reports, pictures and prints circulated, spreading the structure's renown beyond the region and attracting many visitors from far afield. A smaller 1:4 scale replica footbridge was constructed in Wörlitz Park in Germany as early as 1791, which ranks as the first cast iron bridge to be built in continental Europe. Other bridges modelled on the Iron Bridge were erected across the Severn at Buildwas (1796) and Coalport (1818), in Wearmouth in North East England (1796) and in the Scottish village of Craigellachie (1814). In addition to revolutionising bridge design, the Iron Bridge was a stepping stone to a new architectural era defined by covered markets, railway stations, high-rise buildings and London's famous Crystal Palace, since it was now possible to manufacture large iron castings that allowed for wide spans in construction.

In the following years, cracks appeared in the abutments owing to geological movement in the Severn Gorge. The steep slopes had to be shored up by walls to prevent them from slipping. The chorus of acclaim for the new structure was joined by many critical voices, which only fell silent after the bridge survived the devastating flood of 1795 – the only bridge across the Severn to do so as the rising water was able to flow through the open iron structure. Following substantial damage to the stone land arches on the southern bank, they were replaced by wooden arches, which in turn made way for cast iron arches in 1820.

Due to safety concerns, the bridge was closed to vehicular traffic in 1934, the same year it was designated an Ancient Monument. Tolls were collected from pedestrians up until 1950, when ownership of the bridge passed to Shropshire County Council (today Telford & Wrekin Council). Between 1973 and 1975, an inverted reinforced concrete arch was inserted into the riverbed beneath the bridge to counteract the forces that caused the abutments to move towards each other. Their foundations had been eroded as a result of repeated flooding. In addition, the south pier was reinforced with concrete, while the bridge deck was resurfaced with lighter asphalt. In 1986, the bridge and surrounding gorge were designated a UNESCO World Heritage Site ('Ironbridge Gorge').

The original cast iron structure remained largely unchanged over time, apart from being repainted and fitted with metal clamps and bolts to stabilise the dovetails and tenon and mortise joints that had broken as a result of tensile stress. English Heritage, the charity looking after historic buildings and monuments in England into whose guardianship the Iron Bridge had been given, conducted a thorough survey of the structure from 1999 to 2000, followed by extensive research on materials, surfaces, construction techniques and the history of the bridge. A detailed conservation project was conceived in collaboration with Historic England, the public body responsible for championing and protecting the national heritage. The survey had shown that geological movement and an earthquake in the late nineteenth century had caused greater damage than expected to the entire iron structure.

After plans were drafted by Ferguson Mann Architects from Bristol and the London-based engineering firm The Morton Partnership, the contract for the work was awarded to Taziker Industrial, a construction company based in Bolton that specialises in the refurbishment of bridges. The project goals were defined by English Heritage as follows: preserving the original fabric, preventing further loss of historical material, keeping interventions to a minimum, adopting a pragmatic approach and applying tried-and-tested technical solutions to repairing defects. To complement the conservation project, English Heritage launched a public relations and participatory campaign that involved intensive media work, a crowdfunding initiative and a public outreach programme called 'Conservation in Action'. For this purpose, a walkway was built as part of the scaffolding which allowed visitors to watch the craftsmen and conservators going about their work and to listen to the explanations of volunteer guides. As the project's funding partner, the HERMANN REEMTSMA STIFTUNG made a substantial donation, thus placing the project on a secure financial footing and enabling it to go ahead as soon as possible.

Die eingerüstete Iron Bridge, 2018,
mit dem Laufsteg zur Besichtigung der Konservierungsarbeiten des English Heritage /
The Iron Bridge encased in scaffolding, 2018,
with a walkway offering views of the conservation work carried out by English Heritage

The
IRON BRIDGE
TOLLHOUSE

←
Iron Bridge, 2018/19,
Blick von Süden über die Fahrbahn der Brücke auf
den Ort, links das Zollhaus, heute ein Souvenirladen /
view of the town from the south side across the roadway, with
the toll house on the left which today serves as a souvenir shop

Originale Tafel am Zollhaus mit den einzelnen Nutzungstarifen, die Brücke wird als Privateigentum ausgewiesen, von der Zahlung waren selbst Soldaten und die Königliche Familie nicht befreit / Original table of tolls displayed at the toll house pointing out that the bridge was private property and that even soldiers and the Royal Family were not exempt from payment

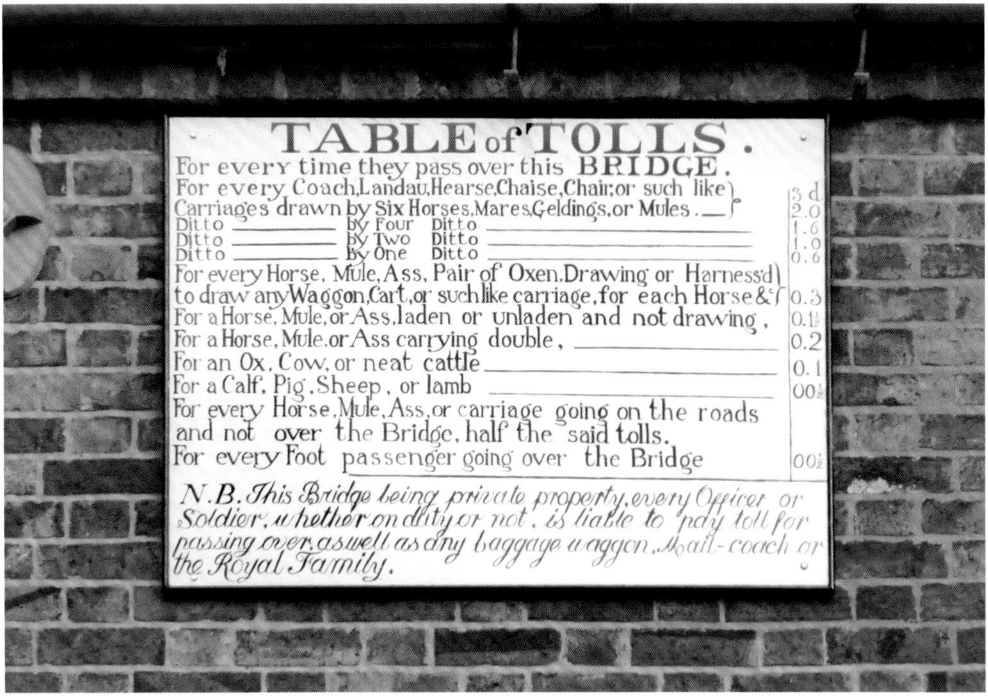

TABLE of TOLLS.

For every time they pass over this BRIDGE.

	s d
For every Coach, Landau, Hearse, Chaise, Chair, or such like Carriages drawn by Six Horses, Mares, Geldings, or Mules.	2.0
Ditto — by Four Ditto	1.6
Ditto — by Two Ditto	1.0
Ditto — by One Ditto	0.6
For every Horse, Mule, Ass, Pair of Oxen, Drawing or Harness'd to draw any Waggon, Cart, or suchlike carriage, for each Horse &c	0.3
For a Horse, Mule, or Ass, laden or unladen and not drawing,	0.1½
For a Horse, Mule, or Ass carrying double,	0.2
For an Ox, Cow, or neat cattle	0.1
For a Calf, Pig, Sheep, or lamb	00½
For every Horse, Mule, Ass, or carriage going on the roads and not over the Bridge, half the said tolls.	
For every Foot passenger going over the Bridge	00½

N.B. This Bridge being private property, every Officer or Soldier, whether on duty or not, is liable to pay toll for passing over, as well as any baggage waggon, Mail-coach or the Royal Family.

The Iron Bridge, 2019,
Ost-Ansicht, Ring und Kielbogen rechts als ornamenthafte Verstrebung, auf den äußeren Spannbögen ist Ort und Errichtungsdatum eingegossen und farblich hervorgehoben / view from the east, with the ornamental circle and ogee serving as braces on the right and the place and year of construction inscribed and painted on the outer rib

Iron Bridge, 2019,
Südwest-Ansicht, hinten der Auflagerbau aus Stein /
view from the south-west, with the stone abutment at the back

Conservation project

The project comprised four areas of work, all carried out between mid-2017 and late 2018:

- Ironwork: securing, repairing, supplementing and in some cases replacing broken parts, especially radials and lateral braces, which were cast anew in Coalbrookdale's blast furnaces using traditional techniques; repairing the cast iron deck plates and their supports; replacing non-original elements (ogee); reversing and reinforcing previous south span repairs; selectively reinforcing the feet of the arches; enhancing the safety of the railings
- Iron surfaces: grit blasting to remove corrosion; applying new protective coating; repainting the entire ironwork based on original findings
- Masonry: repairing the abutments, piers and parapets; replacing stones; repointing the stonework
- Bridge deck resurfacing: removing the cracked and leaking surface; installing a waterproof membrane to protect the iron deck plates; surface water drainage

The Iron Bridge was officially reopened on 1 January 2019, having been repainted in its original red-brown colour. In 2020, the project was awarded the European Heritage Award in the category 'Conservation'. It is open to the public all year round.

Konservierungsprojekt

Vier Bereiche umfassten die Maßnahmen, die von Mitte 2017 bis Ende 2018 durchgeführt wurden:

- Gusseisenstruktur: Sicherung, Reparatur, Ergänzung und fallweise Austausch gerissener Teile, besonders der Radialsprossen und Querstreben, die auf traditionelle Weise in den Hochöfen von Coalbrookdale neu gegossen wurden, Sanierung der gusseisernen Deckplatten und deren Verbindungen, Austausch nicht originaler Teile (ein Kielbogen), Aufhebung und Verbesserung älterer Reparaturen am südlichen Spannbogen, punktuelle Verstärkung der Bogenfüße, Sicherungsarbeiten am Geländer.
- Eisenoberfläche: Entfernung der Korrosionsstellen mittels Sandstrahlung, Auftrag von Korrosionsschutz, Neuanstrich der gesamten Eisenkonstruktion auf Basis von originalen Befunden.
- Mauerwerksarbeiten an den Widerlagern, Pfeilern und der Brüstung, Austausch von Steinen, Neuverfugungen.
- Erneuerung der Fahrbahn: Entfernung des rissigen und undichten Belags, Einbringung einer wasserdichten Schicht als Korrosionsschutz der Eisenplatten, Ableitung des Oberflächenwassers.

Am 1. Januar 2019 wurde die Iron Bridge feierlich wieder eingeweiht. Sie hatte ihre ursprüngliche rotbraune Farbe zurückerhalten. 2020 wurde das Projekt mit dem »European Heritage Award« in der Kategorie »Conservation« ausgezeichnet. Die Brücke ist ganzjährig öffentlich zugänglich.

Iron Bridge mit Zollhaus im Hintergrund, 2019, Nordost-Ansicht,
der rotbraune Anstrich entspricht originalen Befunden
und dem historischen Gemälde von William Williams von 1780 (Abb. S. 16) /
The Iron Bridge with the toll house in the background, 2019, view from the north-east,
the red-brown coating corresponds to original findings
and to the historical painting by William Williams from 1780 (fig. p. 16)